RAILWAY INDUSTRY STANDARD
OF THE PEOPLE'S REPUBLIC OF CHINA

Alloy Steel Built-up Crossings

TB/T 3467-2016

Issued by National Railway Administration of the People's Republic of China
Issued Date: December 15, 2016
Valid Date: July 1, 2017

China Railway Publishing House Co., Ltd.
Beijing 2021

图书在版编目(CIP)数据

合金钢组合辙叉:TB/T 3467—2016:英文/中华人民共和国国家铁路局组织编译. —北京:中国铁道出版社有限公司,2021. 4
ISBN 978-7-113-27568-6

Ⅰ. ①合… Ⅱ. ①中… Ⅲ. ①合金钢-辙叉-行业标准-中国-英文 Ⅳ. ①TG142. 33-65

中国版本图书馆 CIP 数据核字(2020)第 268006 号

Chinese version first published in the People's Republic of China in 2017
English version first published in the People's Republic of China in 2020
by China Railway Publishing House Co., Ltd.
No. 8, You'anmen West Street, Xicheng District
Beijing, 100054
www. tdpress. com

Printed in China by BEIJING CONGREAT PRINTING CO.,LTD.

ISBN 978-7-113-27568-6

Introduction to the English Version

To promote the exchange and cooperation in railway technology between China and the rest of the world, Planning and Standard Research Institute of National Railway Administration, entrusted by National Railway Administration of the People's Republic of China, organized the translation and preparation of Chinese railway technology and product standards.

This Standard is the official English version of *Alloy Steel Built-up Crossings* (TB/T 3467-2016). The original Chinese version of this standard was issued by National Railway Administration of the People's Republic of China and came into effect on July 1, 2017. In case of discrepancies between the two versions, the Chinese version shall prevail. National Railway Administration of the People's Republic of China owns the copyright of this English version.

Your comments are invited for next revision of this standard and should be addressed to Technology and Legislation Department of National Railway Administration and Planning and Standard Research Institute of National Railway Administration.

Address: Technology and Legislation Department of National Railway Administration, No. 6, Fuxing Road, Beijing, 100891, P. R. China.

Planning and Standard Research Institute of National Railway Administration, B block, JianGong Building, No. 1, Guanglian Road, Xicheng District, Beijing, 100055, P. R. China.

The translation was performed by Wang Lei.

The translation was reviewed by Gu Aijun, Chen Qi, Si Daolin, Han Zhigang, Dai Ruoyu, Xue Jigang and Zhao Lin.

Notice of National Railway Administration on Issuing the English Version of Eight Chinese Railway Technical Standards including *Strength Design and Test of Multiple Unit Body Structures*

Document GTKF 〔2020〕 No. 41

Eight Chinese railway technical standards including *Strength Design and Test of Multiple Unit Body Structures* (TB/T 3451-2016) are issued. In case of discrepancies between the two versions, the Chinese version shall prevail.

China Railway Publishing House Co., Ltd. is authorized to publish and distribute the English version of these standards.

Attached here is a list of the English version of these standards.

S/N	Chinese title	English title	Standard number
1	动车组车体结构强度设计及试验	Strength Design and Test of Multiple Unit Body Structures	TB/T 3451-2016
2	列车牵引计算 第1部分:机车牵引式列车	Railway Train Traction Calculation— Part 1: Trains with Locomotives	TB/T 1407.1-2018
3	机车车辆强度设计及试验鉴定规范 车体 第1部分:客车车体	Strength Design and Test Accreditation Specification for Rolling Stock—Car Body— Part 1: Passenger Car Bodies	TB/T 3550.1-2019
4	机车车辆强度设计及试验鉴定规范 车体 第2部分:货车车体	Strength Design and Test Accreditation Specification for Rolling Stock—Car Body— Part 2: Freight Car Bodies	TB/T 3550.2-2019
5	铁路信号故障—安全原则	Fail-safe Principle of Railway Signaling	TB/T 2615-2018
6	高速铁路预制先张法预应力混凝土简支梁	Precast Pretensioned Prestressed Concrete Simple-supported Beam of High-speed Railway	TB/T 3433-2016
7	合金钢组合辙叉	Alloy Steel Built-up Crossings	TB/T 3467-2016
8	铁路电力低电阻接地系统成套装置	Low Resistance Grounding Assembly for Electric Power System of Railway	TB/T 3465-2016

National Railway Administration of the People's Republic of China

October 27, 2020

Contents

Foreword ·· Ⅲ
1 Scope ·· 1
2 Normative References ·· 1
3 Technical Requirements ·· 2
4 Factory Assembly ·· 5
5 Inspection Method ·· 8
6 Inspection Rules ·· 11
7 Marks, Quality Certificates, Storage and Transport ·· 11
Annex A(Normative) Basic Items for Crossing Inspection ·· 13

Foreword

This Standard is drafted in accordance with the rules given in GB/T 1.1-2009.

This Standard is proposed and managed by China Railway Economic and Planning Research Institute.

This Standard is prepared by China Railway Engineering Consulting Group Co., Ltd., Metals & Chemistry Research Institute of China Academy of Railway Sciences, Railway Engineering Research Institute of China Academy of Railway Sciences, Zhejiang Desheng Railway Equipment Co., Ltd., Sichuan Blue Star Machinery Co., Ltd., and Beijing Teye Industrial and Trade Corp.

This Standard is mainly drafted by Xu Youquan, Luo Yan, Qiao Shenlu, Zhao Tianyun, Zhang Dongfeng, He Xuefeng, Zhou Qingyue, Zhang Yinhua, Zou Dingqiang, Li Lianxiu, Wang Shuguo, Guan Tie, Li Bennan and Zhang Zhiqiang.

Alloy Steel Built-up Crossings

1 Scope

This Standard specifies the technical requirements, factory assembly, inspection methods, inspection rules, marks, storage, transport and quality certificates of alloy steel built-up crossings.

This Standard is applicable to the alloy steel built-up crossings (hereinafter referred to as crossings) through which a train passes at the speed of less than or equal to 160 km/h.

2 Normative References

The following referenced documents are indispensable for the application of this Standard. For dated references, only the edition cited applies. For undated references, the latest edition of the referenced document (including any amendments) applies.

GB/T 228.1 *Metallic Materials—Tensile Testing—Part 1: Method of Test at Room Temperature*

GB/T 229 *Metallic Materials—Charpy Pendulum Impact Test Method*

GB/T 230.1 *Metallic Materials—Rockwell Hardness Test—Part 1: Test Method (Scales A, B, C, D, E, F, G, H, K, N, T)*

GB/T 231.1 *Metallic Materials—Brinell Hardness Test—Part 1: Test Method*

GB/T 1228 *High Strength Bolts with Large Hexagon Head for Steel Structures*

GB/T 1229 *High Strength Large Hexagon Nuts for Steel Structures*

GB/T 1230 *High Strength Plain Washers for Steel Structures*

GB/T 1231 *Specifications of High Strength Bolts with Large Hexagon Head, Large Hexagon Nuts, Plain Washers for Steel Structures*

GB/T 2828.1 *Sampling Procedures for Inspection by Attributes—Part 1: Sampling Schemes Indexed by Acceptance Quality Limit (AQL) for Lot-by-lot Inspection*

GB/T 6402 *Steel Forgings—Method for Ultrasonic Testing*

GB/T 6414 *Castings—System of Dimensional Tolerances and Machining Allowances*

GB/T 6461-2002 *Methods for Corrosion Testing of Metallic and Other Inorganic Coatings on Metallic Substrates—Rating of Test Specimens and Manufactured Articles subjected to Corrosion Tests*

GB/T 11352 *Carbon Steel Castings for General Engineering Purpose*

TB/T 412-2014 *Technical Specification on Turnouts for Standard-gauge Railway*

TB/T 1495.2 *Spring Clip-I Fastenings—Spring Clip*

TB/T 1632.2-2014 *Welding of Rails—Part 2: Flash Butt Welding*

TB/T 2344-2012 *Technical Specifications for the Procurement of 43 kg/m-75 kg/m Rails*

TB/T 2975 *Specification for Bonded Insulated Rail Joints Used on Railway*

TB/T 3065.2 *Spring Clip-II Fastenings—Part 2: Spring Clip*

TB/T 3109-2013 *Asymmetric Cross Section Rails for Railway Turnouts*

TB/T 3307.2-2014 *Technical Specification for Manufacturing of High Speed Turnouts—Part 2: T-bolts*

TB/T 3307.4-2014 *Technical Specification for Manufacturing of High Speed Turnouts—Part 4: Gauge Blocks*

3 Technical Requirements

3.1 General Requirements

3.1.1 The crossings shall be manufactured in accordance with the design drawings approved under specified procedures and this Standard.

3.1.2 The crossings may be classified into the following four types: forged alloy steel built-up point crossing, alloy steel rail built-up crossing, alloy steel built-up crossing inlaid with wing rail (the point and wing rail insert may be an integral structure) and reinforced alloy steel built-up crossing welded with wing rail, as respectively shown in a), b), c) and d) in Figure 1. Glue may be used to bond the spacing block of crossings and the rails.

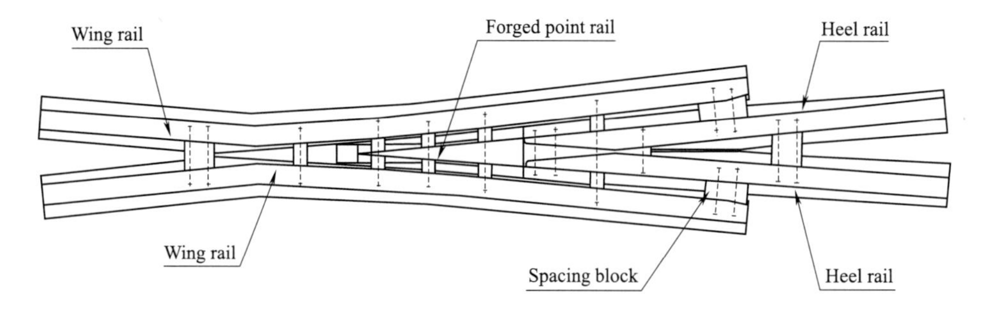

a) Forged Alloy Steel Built-up Point Crossing

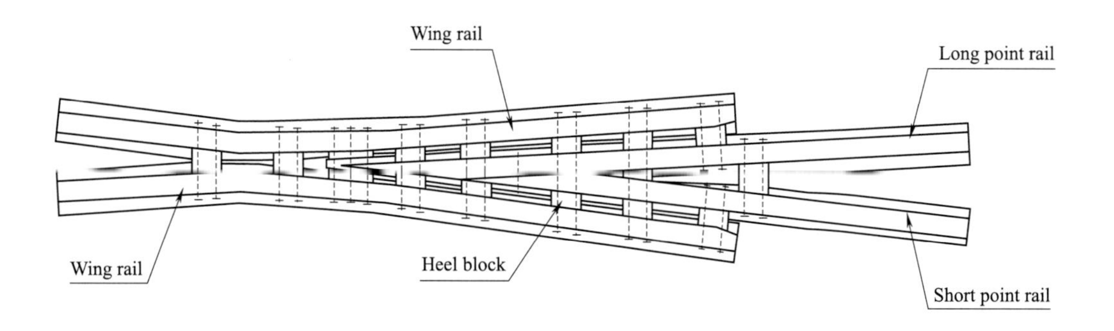

b) Alloy Steel Rail Built-up Crossing

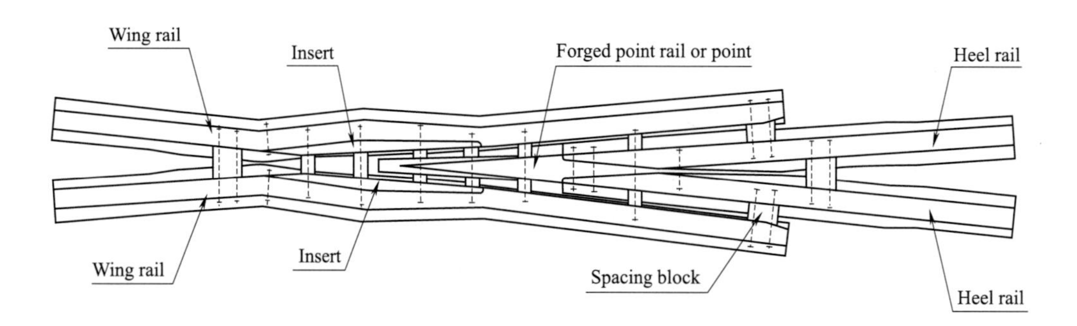

c) Alloy Steel Built-up Crossing Inlaid with Wing Rail

Figure 1 Alloy Steel Built-up Crossings

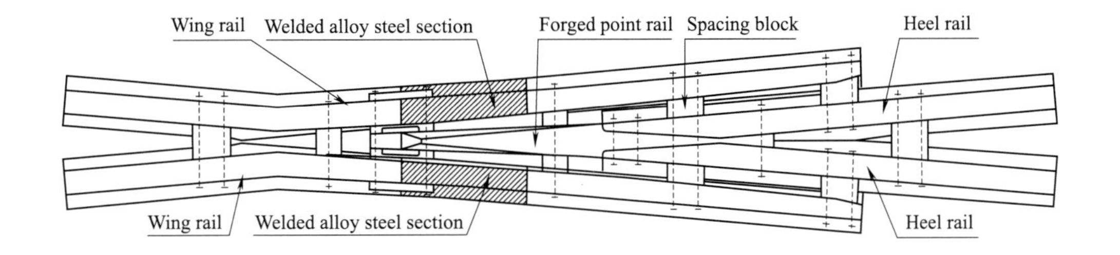

d) **Reinforced Alloy Steel Built-up Crossing Welded with Wing Rail**

Figure 1 Alloy Steel Built-up Crossings(continued)

3.2 Raw Materials

3.2.1 The mechanical properties of alloy steel components shall comply with Table 1.

Table 1 Mechanical Properties of Alloy Steel Components for Crossings

Item	Tensile strength R_m MPa	Elongation at break A	Reduction rate of cross section Z	Absorbed impact energy on rail head KU_2 J	Hardness	
					Rail top surface HBW 10/3000	Cross section HRC
Alloy steel rail	≥1 280	≥12%	—	Normal temperature (20 ℃)≥60; Low temperature (−40 ℃)≥30	360-430	38-45
Forged point rail, point, wing rail insert, alloy steel section of welded wing rail	≥1 280	≥12%	≥40%	Normal temperature (20 ℃)≥60; Low temperature (−40 ℃)≥30		

3.2.2 The content of sulfur, phosphorus and hydrogen in alloy steel shall not exceed 0.015%, 0.025% and 0.000 15% respectively.

3.2.3 No crack is allowed in the heat-treated area.

3.2.4 The non-alloy steel rail used for manufacturing crossings shall undergo on-line heat treatment and comply with TB/T 2344-2012.

3.3 Rail components and parts

3.3.1 The inclination of the rail end face in the vertical and horizontal direction shall be less than 1.0 mm.

3.3.2 The tolerance for the lengths of long point rail, short point rail and heel rail shall be ±2 mm. The tolerance for the width of rail base shall be within −2 mm-0 mm and that of rail head (except for the bending point) shall be ±0.5 mm.

3.3.3 The tolerance for the length of wing rail shall be ±6 mm and that for the height shall be ±0.5 mm. The tolerance for the width of rail base shall be within −2 mm-0 mm and that for the rail head (except for the bending point) shall be ±0.5 mm.

3.3.4 The tolerance for the height of each control profile in the rail head machining section shall be ±0.5 mm.

3.3.5 Where the rail head widths of the long and short point rails are 35 mm, 50 mm, 71 mm (or 70 mm), and the tolerance for the profile of the rail head in rail profile shall not exceed 0.3 mm.

3.3.6 Rails prone to cracks after being bended and twisted shall not be used for the crossings. During local heating and bending, the alloy steel rail should be bent at the normalizing temperature and the heating temperature for the non-alloy steel rail shall not be greater than 500 ℃. The indentation depth incurred in bending shall not be greater than 0.5 mm.

3.3.7 The tolerance of rail cant shall be ±1 : 320.

3.3.8 The edges and corners of the binding surface of rail head, point rail and insert etc. shall be chamfered to 1.5 mm×45°-2.5 mm×45°. The rail end face and the edges and corners of machined rail head and rail base shall be chamfered to 1.0 mm×45°-2.0 mm×45° or rounded.

3.3.9 The machined surface of rails shall be smooth, and the surface roughness shall be MRR *Ra*25 μm.

3.4 Forged Point Rail

3.4.1 The tolerance of the length from the actual point of crossing to the starting point of binding surface of the heel rail shall be ±2 mm.

3.4.2 The tolerance for the width of each cross section shall be ±0.5 mm.

3.4.3 Where the width of the point rail head is in the range of 20 mm-50 mm, the tolerance for each of the cross section shall be ±0.5 mm.

3.4.4 The straight-line point rail shall be symmetrical to the centerline, and the asymmetry of the machined surfaces in contact shall not exceed 0.5 mm.

3.4.5 The point rail shall not be rectified in thermal state after the forging and heat treatment.

3.4.6 The edges and corners of the binding surface of point rail and heel rail shall be chamfered to 1.5 mm×45°-2.5 mm×45°, and the cutting face at the rail top of actual point of crossing shall be rounded off.

3.4.7 No slag, crack and other defects are allowed inside the point rail.

3.4.8 The external top surface of the point rail shall be free from oxide scale. The area of oxide scale on the bottom surface, if any, shall not exceed 1 mm×50 mm (depth × length) and that on the side surface shall not exceed 0.5 mm×50 mm (depth × length).

3.4.9 For the point with partial wing rails, the width deviation of the flangeway shall comply with Article 4.5.5 and the depth of the flangeway shall comply with Article 4.5.9.

3.5 Wing Rail Insert

3.5.1 The tolerance for the length of the wing rail insert shall be within −2 mm-0 mm. The tolerance for the width (*B*) of each control section shall be ±0.5 mm, the tolerance of height (*H*) shall be within 0 mm-+1 mm and that of thickness (*A*) shall be ±0.5 mm (see dimensions *A*, *B* and *H* in Figure 2). The edges and corners of the binding surfaces of insert and wing rail head shall be chamfered to 1.5 mm×45°-2.5 mm×45°.

3.5.2 No slag, crack and other defects are allowed inside the wing rail insert.

3.5.3 The external top surface of the wing rail insert shall be free from oxide scale. The area of oxide scale on the bottom surface, if any, shall not exceed 1 mm×50 mm (depth × length) and that on the side surface shall not exceed 0.5 mm×50 mm (depth × length).

3.6 Welded Wing Rail

The flash butt welding of wing rail shall comply with TB/T 1632.2-2014 in aspects such as appearance of welds, ultrasonic flaw detection, drop hammer, static bending, fatigue, hardness, impact, fracture, tensile strength (R_m) and elongation at break (*A*) of welded joints.

3.7 Bolt Holes of Point Rail, Rail and Insert

The tolerance of bolt holes shall comply with the following requirements:

Unit: mm

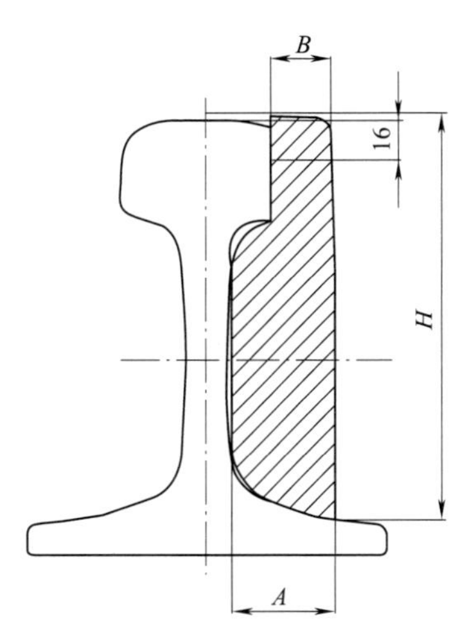

Figure 2 Dimensions of Wing Rail Insert

a) The tolerance of hole diameter shall be 0 mm-+1.0 mm.

b) The tolerance for the upper and lower positions of the hole center shall be ±1.0 mm.

c) The tolerance for the distance between centers of holes to be joined and between two adjacent holes shall be ±1.0 mm.

d) The tolerance of the distance between holes not to be joined shall be ±2.0 mm and the tolerance of the center to center distance between two holes with the furthest distance shall be ±3.0 mm.

e) The tolerance of the distance from the center of the joint bolt hole to the rail end shall be ±1.0 mm.

f) The bolt hole roughness shall be MRR *Ra*25 μm. The bolt holes shall be chamfered to 0.8 mm×45°-1.5 mm×45° and the burrs on the bolt holes shall be removed.

3.8 Spacing Block

3.8.1 The materials of spacing block shall comply with the design requirements and GB/T 11352.

3.8.2 The tolerance for the fitting dimensions of rail components and parts, wing rail insert and point rail shall comply with the design requirements. The tolerance of other dimensions shall comply with GB/T 6414.

3.9 Base Plate

3.9.1 The welding and dimensional deviation of base plate shall comply with Article 3.4.1 of TB/T 412-2014.

3.9.2 The tolerance of the distance from the gauge blocks mounted on both sides of the base plate to the running surface shall be ±1.0 mm.

3.9.3 The tolerance of the distance between two iron bases on the base plate shall be ±1.0 mm.

3.9.4 The tolerance for the parallelism of two parallel iron bases on the base plate shall be 1.0 mm.

3.9.5 The bolt holes on the base plate shall be chamfered to1.0 mm×45°-1.5 mm×45°.

4 Factory Assembly

4.1 General Requirements

4.1.1 The built-in crossings shall not be shipped out of factory until they have passed quality inspection.

4.1.2 The high-strength bolt sets of Grade 10.9 or above shall be adopted for the assembly of crossings and such bolt sets shall comply with the design requirements as well as GB/T 1228, GB/T 1229, GB/T 1230 and GB/T 1231.

4.1.3 The high-strength bolts shall be tightened at 100%-110% of the design torque, and shall be re-tightened before delivery.

4.1.4 The elastic pads under iron base plate and rail may be assembled with the crossings or may be supplied separately according to the requirements of users.

4.1.5 The narrow gap between the spacing block and the point rail, wing rail and heel rail which are not glued shall be less than 0.5 mm. The rails and spacing blocks shall be glued in accordance with the following requirements:

a) Burrs on the bolt holes and gluing surface of rails and spacing blocks shall be removed, and the protruding marks on the side of rail web shall be ground smooth.

b) The gluing surface of rails and spacing blocks shall be sandblasted, after which the gluing surface shall be free from rust spot, oil stain and moisture.

c) The properties of the material in the glued layer shall comply with TB/T 2975.

4.1.6 The spring clips, T-bolts, gauge blocks and elastic pads used for the assembly of crossings shall comply with the following requirements:

a) The spring clip-I shall comply with TB/T 1495.2 and the spring clip-II shall comply with TB/T 3065.2.

b) The T-bolts shall comply with TB/T 3307.2-2014.

c) The gauge blocks shall comply with TB/T 3307.4-2014.

d) The elastic pads shall comply with TB/T 412-2014.

4.2 Assembly of Forged Alloy Steel Built-up Point Crossing and Reinforced Alloy Steel Built-up Crossing Welded with Wing Rail

4.2.1 The point rail and heel rail shall be first assembled into a subassembly, and then with the wing rail.

4.2.2 The binding surface gap between the heel rail head and the point rail shall be less than 0.5 mm.

4.2.3 The top surface and running surface of the heel rail shall not protrude from the top surface and running surface of the point rail.

4.3 Assembly of Alloy Steel Rail Built-up Crossing

4.3.1 The long and short point rails shall be first assembled into a subassembly, and then with the wing rail.

4.3.2 The tolerance for the offset of long and short point rails shall be ±1.0 mm.

4.3.3 The gap between the wing rail base and the point rail base shall be greater than or equal to 2.0 mm.

4.3.4 The binding gap between the long and short point rail heads shall be less than 0.5 mm. The gap between the long and short point rail bases shall be greater than or equal to 1.0 mm (see dimensions S and b in Figure 3).

4.4 Assembly of Alloy Steel Built-up Crossing Inlaid with Wing Rail

4.4.1 The wing rail and insert as well as the forged point rail and heel rail shall be respectively assembled into subassemblies, and then into a whole. To assemble the alloy steel crossing integrating the point and wing rail insert, the wing rail and the integral point as well as the integral point and the heel rail shall be respectively assembled into subassemblies, and then into a whole.

4.4.2 After the assembly, the gap between the wing rail head and the insert as well as that between the point rail and the heel rail shall be less than 0.5 mm.

4.4.3 The gap between the insert and the wing rail base shall be less than 0.5 mm and the tolerance

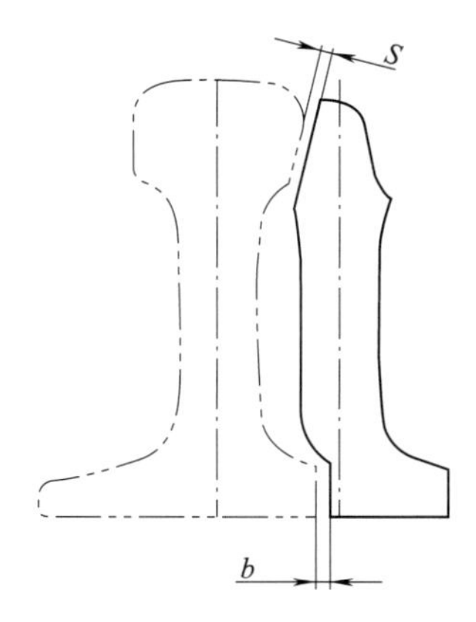

Figure 3　Gaps between Long and Short Point Rails

of the height difference (h) between each control section at the front end of the insert and the top surface of the wing rail shall be within -1.0 mm-0 mm. The back end of the insert shall not protrude from the wing rail top (see dimension h in Figure 4).

Unit: mm

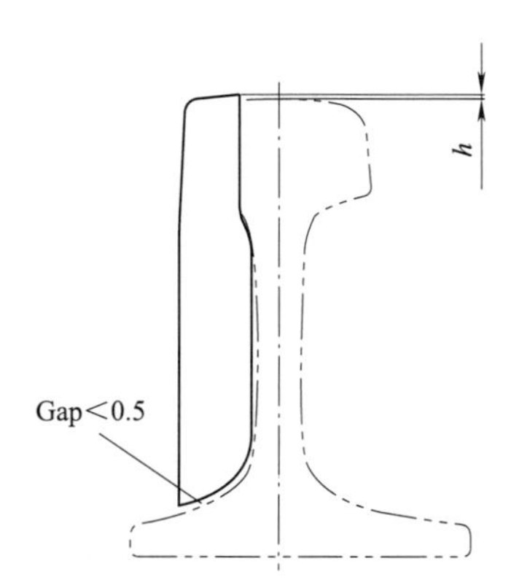

Figure 4　Gap between Insert and Wing Rail

4.5　Tolerance of Assembled Crossing

4.5.1　The tolerance for the total length of crossings shall be ±4 mm, as shown in Figure 5.

Unit: mm

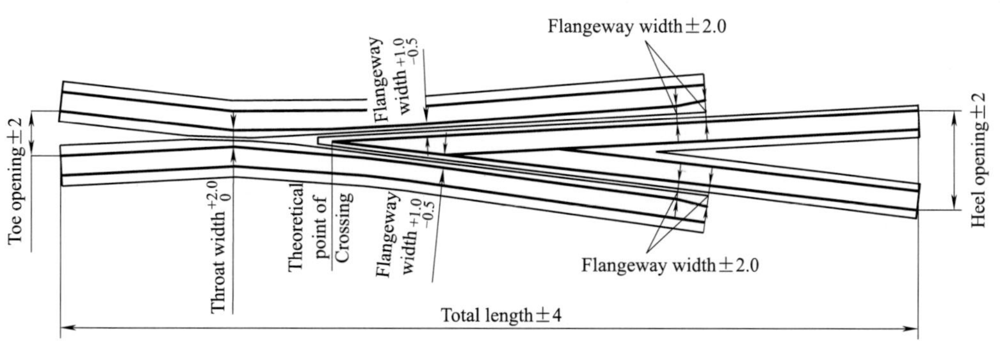

Figure 5　Tolerance of Assembled Crossing

4.5.2　The tolerance of the length from the crossing toe end and heel end to the theoretical point of crossing shall be ±2 mm.

4.5.3　The tolerance of the toe opening and heel opening shall be ±2 mm.

4.5.4　The tolerance of the throat width shall be within 0 mm-+2.0 mm.

4.5.5 For the point rail with a rail head width of 20 mm and 50 mm, the tolerance of the flangeway width in the cross section shall be −0.5 mm-+1.0 mm and that in the buffer section at the back end of wing rail shall be ±2.0 mm.

4.5.6 For the point rail with a rail head width of 20 mm and 50 mm (or 45 mm), the tolerance of the height (*h*) from the top of point rail to that of wing rail (or insert) in the cross section shall be ±0.5 mm (see dimension *h* in Figure 6).

Unit: mm

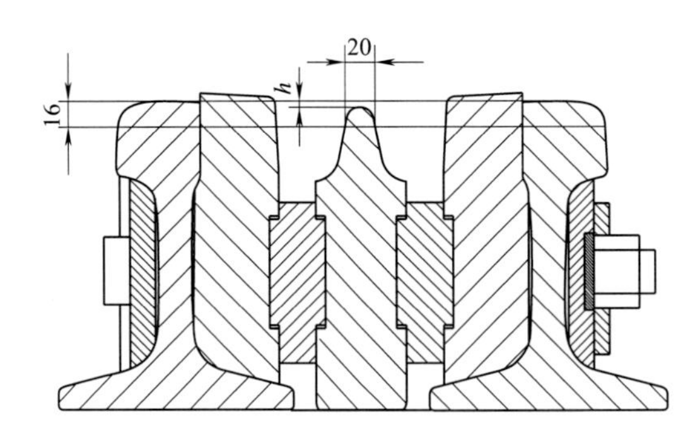

Figure 6 Tolerance of Crossing Assembly Section

4.5.7 The tolerance for the offset of each control section of the curved crossing shall be ±1.0 mm.

4.5.8 The straightness of the running surface of crossings shall be 2.0 mm and there shall be no anti line within the section area of 50 mm from the throat to the point rail. The running surface of curved crossings shall be round and smooth.

4.5.9 The flangeway depth shall not be less than 47 mm.

4.5.10 The gap between the point rail base and the wing rail base shall not be less than 2.0 mm.

4.5.11 When the point rail base is flush with the wing rail base, the flatness of the point rail base and the wing rail base within the contact area with the base plate shall be 1.0 mm.

4.5.12 After the iron base plate is installed for the crossings, the gap between the wing rail and point rail base, and the elastic pad or iron base plate shall be less than 0.5 mm.

4.5.13 For the crossings with a rail cant, the tolerance of its rail cant shall be 1 : 320.

4.5.14 The complete set of assembled crossings shall also comply with Annex A in addition to the above requirements.

5 Inspection Method

5.1 Alloy Steel Rail

5.1.1 Tensile Property

The tensile property shall be inspected according to GB/T 228.1. For the rail with symmetrical section, the tensile samples shall be taken at the positions specified in TB/T 2344-2012; for the rail with asymmetric section, the tensile samples shall be taken at the positions specified in TB/T 3109-2013.

5.1.2 Impact Resistance

The impact resistance shall be tested according to GB/T 229. Four impact test samples shall be taken by cutting 2-mm-deep U-shaped notches facing the rail side at the positions on the rail head shown in Figure 7. The impact test results of the four samples shall be averaged.

Unit: mm

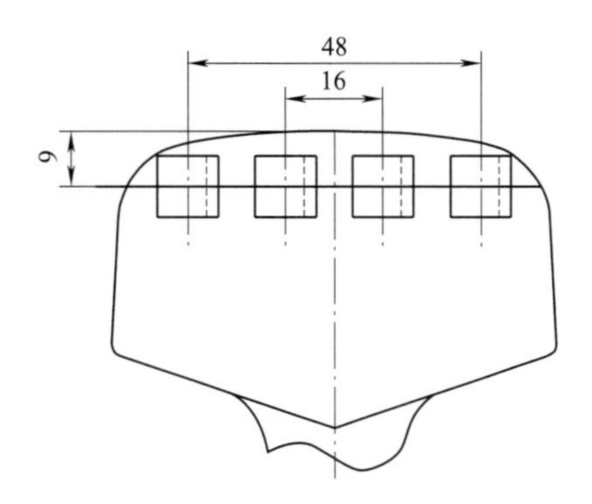

Figure 7 Sampling for Rail Impact Test

5.1.3 Hardness

5.1.3.1 Rail Top Surface Hardness

The hardness shall be tested according to the method specified in GB/T 231.1. During the process, the top surface of rail head shall be ground off for 0.5 mm and the Brinell hardness shall be tested along the centerline of rail head. The sample length shall not be less than 100 mm and the number of test points shall not be less than 5. The test results shall be averaged as the final value.

5.1.3.2 Cross Section Hardness

The cross section hardness shall be tested according to the method specified in GB/T 230.1. For the rail with symmetrical section, the Rockwell hardness of cross section of the rail shall be measured at the measuring points shown in Figure 8a); for the rail with asymmetric section, the Rockwell hardness of cross section of the rail shall be measured at the measuring points shown in Figure 8b).

5.1.4 Ultrasonic Flaw Detection

The ultrasonic flaw detection shall be performed according to the method specified in Section 7.7 of TB/T 2344-2012.

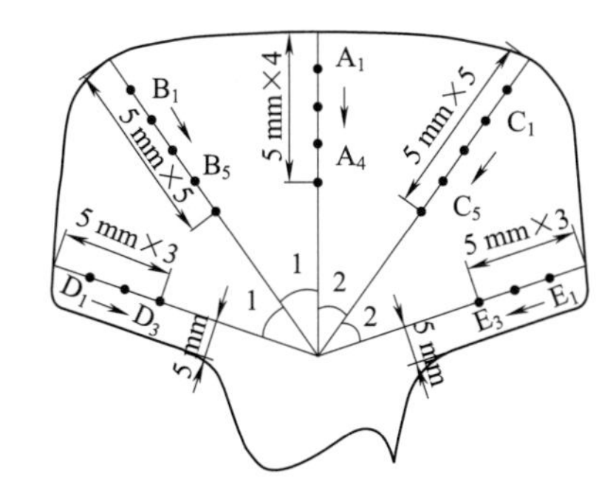

a) Test Positions of Cross Section Hardness of Rail With Symmetrical Section

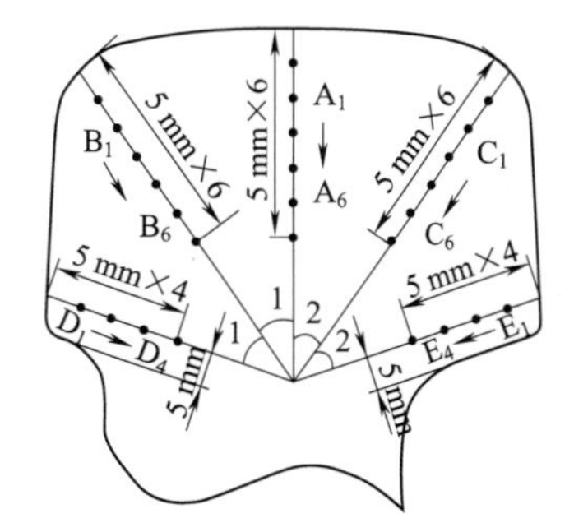

b) Test Positions of Cross Section Hardness of Rail With Asymmetric Section

Note: The first point is 5 mm away from the surface and all the other points are distributed at a spacing of 5 mm. For example, 5 mm × 3 means 3 points at a spacing of 5 mm in between them. The distance between lines D, E and the jaw is 5 mm. Lines B and C are angular bisectors of lines A and D as well as lines A and E.

Figure 8 Test Positions of Cross Section Hardness of Rail

5.1.5 Content of Sulfur, Phosphorus and Hydrogen

Sampling and inspection shall be conducted according to the method specified in TB/T 2344-2012.

5.2 Forged Point Rail, Point, Wing Rail Insert, and Alloy Steel Section of Welded Wing Rail

5.2.1 Tensile Property

The tensile property shall be inspected according to GB/T 228.1 and d_0 is 10 mm. The tensile samples shall be taken at the positions shown in Figure 9a) and the standard distance of samples (L_0) is $5d_0$ before the application of force at room temperature.

5.2.2 Impact Resistance

The impact resistance shall be tested according to GB/T 229, and the samples shall be taken by cutting 2-mm-deep U-shaped notches at the positions shown in Figure 9 a).

5.2.3 Hardness

5.2.3.1 Top Surface Hardness

The top surface hardness shall be tested according to GB/T 231.1. Three points shall be sampled respectively in the two cross sections of the point rail with a width of 40 mm and 70 mm, and the test results shall be averaged as the final value; three points shall be sampled at the positions where the insert widths are 35 mm and 66 mm respectively, and the test results shall be averaged as the final value.

5.2.3.2 Cross Section Hardness

The cross section hardness shall be inspected according to GB/T 230.1. Three points shall be sampled at each of the two positions, as shown in Figure 9 b), and the test results shall be averaged as the final value.

Unit: mm

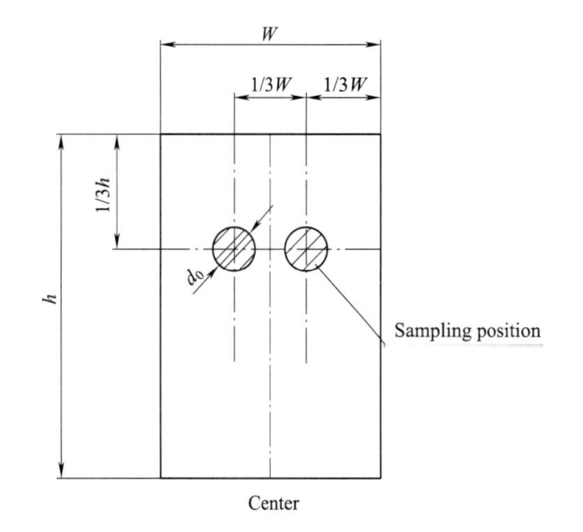

a) Sampling for Tensile and Impact Test Samples

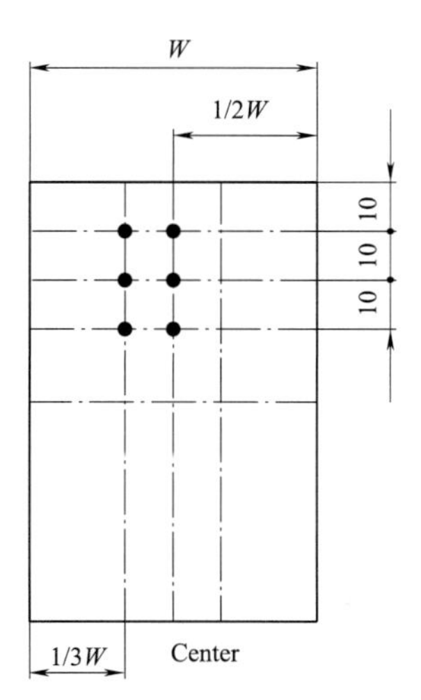

b) Sampling for Cross Section Hardness Test

Figure 9 Sampling Positions of Forged Point Rail, Point, Wing Rail Insert, and Alloy Steel Section of Welded Wing Rail

5.2.4 Ultrasonic Flaw Detection

The straight beam probe shall be used for the one-by-one ultrasonic flaw detection in accordance with GB/T 6402 and the probe frequency shall be 2.5 MHz-5 MHz. The DAC curve used for quantitative analysis of defects (discontinuous) shall be generated with $\phi 3$ mm and $\phi 2$ mm flat-bottomed holes, and the acceptance standard shall conform to quality grade 4.

5.2.5 Contents of Sulfur, Phosphorus and Hydrogen

The contents of sulfur, phosphorus and hydrogen shall be tested according to the method specified in TB/T 2344-2012 and samples shall be taken from the core of parent.

5.3 Machining and Assembly of Crossings

5.3.1 The shape, appearance and dimensions of a complete set of crossings shall be inspected with tape measure, straightedge and chord.

5.3.2 The gap of various crossing components and parts shall be inspected with special template, caliper and feeler gauge.

5.3.3 The length of each part of crossings and the rail head profile shall be inspected with tape measure, straightedge, caliper and special template.

5.3.4 The torque of the bolt sets shall be inspected with a torque wrench.

5.4 Welds of Welded Wing Rail

The welds of welded wing rail shall be inspected according to the method specified in Chapter 4 of TB/T 1632.2-2014.

5.5 Gluing Materials

The epoxy glue cured at room temperature shall be used and the test methods for its physical and mechanical properties shall comply with TB/T 2975.

6 Inspection Rules

6.1 Types of Inspection

The inspection of crossings includes type test and delivery inspection.

6.2 Type Test

6.2.1 The type test of crossings shall be carried out and the test results shall be archived for future reference if one of the following conditions occurs:

a) Production for the first time;

b) Great changes in materials or technologies;

c) Normal production for two years;

d) Resumption of Production after suspension for half a year or more.

6.2.2 The type test items of alloy steel rail, forged point rail, wing rail insert and alloy steel section of welded wing rail shall comply with Annex A.

6.2.3 The type test of welds of the welded wing rail shall comply with Section 5.2 of TB/T 1632.2-2014.

6.3 Delivery Inspection

6.3.1 The crossing components and the assembled crossings shall be inspected by the technical inspection department of the manufacturer and shall not be shipped out of factory until they have passed the inspection.

6.3.2 The delivery inspection shall comply with Annex A.

6.3.3 In the delivery inspection of complete sets of crossings, the products will be judged as qualified if the pass rate of category A items is 100%, the pass rate of category B items is greater than or equal to 90% and that of category C items is greater than or equal to 80%. In the calculation of the pass rate, if the number of inspection points for one specific item is more than one, the pass rate shall be calculated on the basis of multiple inspection points. The crossings will be judged as unqualified if the deviation of a single item of category B is greater than twice the limit and that of a single item of category C is greater than 3 times the limit.

7 Marks, Quality Certificates, Storage and Transport

7.1 Marks

7.1.1 The complete set of crossings shall be labeled with permanent marks, which shall include the crossing type, left hand or right hand type, serial number and date of production, manufacturer name or logo, etc.

7.1.2 The hoisting positions shall be marked on the crossing assembly.

7.1.3 The measuring position of machined reduced depth shall be marked on the wing rail and point rail.

7.2 Quality Certificates

Each set of crossings shall be provided with the installation drawing and product certificate upon delivery.

7.3 Storage and Transport

7.3.1 The ground for stacking the crossing components shall be flat. The number of stacking layers shall not be more than 4 and each layer shall be placed on top of firmly laid wooden blocks with an area of not less than 60 mm × 60 mm for each. Such blocks and their supports shall be arranged vertically along the height.

7.3.2 Collision and falling shall be prevented during the transportation and handling of the crossings.

Annex A
(Normative)
Basic Items for Crossing Inspection

The type test and delivery inspection shall be conducted for the crossings according to the items listed in Table A. 1.

Table A Basic Items for Crossing Inspection

<table>
<tr><th>S/N</th><th colspan="3">Inspection item</th><th>Requirements</th><th>Frequency</th><th>Classification of inspection item</th><th>Type test</th><th>Delivery inspection</th></tr>
<tr><td>1</td><td rowspan="8">Rail components and parts</td><td rowspan="3">Alloy steel rail, forged point rail, point, wing rail insert, and alloy steel section of welded wing rail</td><td>Tensile property</td><td>≥1 280 MPa. If unqualified, double sampling and re-inspection shall be required</td><td>1 for each heat treatment furnace</td><td>B</td><td>√</td><td></td></tr>
<tr><td>2</td><td>Impact resistance</td><td>Normal temperature (20 ℃)≥60 J;
Low temperature (−40 ℃)≥30 J.
If unqualified, double sampling and re-inspection shall be required</td><td>1 for each heat treatment furnace</td><td>B</td><td>√</td><td></td></tr>
<tr><td>3</td><td>Elongation at break</td><td>≥12%. If unqualified, double sampling and re-inspection shall be required</td><td>1 for each heat treatment furnace</td><td>B</td><td>√</td><td></td></tr>
<tr><td>4</td><td colspan="2">Reduction of cross section of forged point rail, point, wing rail insert, alloy steel section of welded wing rail</td><td>≥40%</td><td>1 for each heat treatment furnace</td><td>B</td><td>√</td><td></td></tr>
<tr><td>5</td><td rowspan="2">Alloy steel rail</td><td>Rail top hardness</td><td>HBW 360-HBW 430</td><td>Type test: 1 for each heat treatment furnace Delivery inspection: one by one</td><td>A</td><td>√</td><td>√</td></tr>
<tr><td>6</td><td>Cross section hardness</td><td>HRC 38-HRC 45</td><td>1 for each heat treatment furnace</td><td>A</td><td>√</td><td>√</td></tr>
<tr><td>7</td><td rowspan="2">Forged point rail, point, wing rail insert, alloy steel section of welded wing rail</td><td>Rail top hardness</td><td>HBW 360-HBW 430</td><td rowspan="2">One by one</td><td rowspan="2">A</td><td rowspan="2">√</td><td rowspan="2">√</td></tr>
<tr><td>8</td><td>Cross section hardness</td><td>HRC 38-HRC 45. If unqualified, parent samples shall be taken for the core hardness test; if it is still unqualified, the product will be judged as unqualified</td></tr>
</table>

Table A Basic Items for Crossing Inspection(continued)

S/N	Inspection item			Requirements	Frequency	Classification of inspection item	Type test	Delivery inspection
9	Rail components and parts	Ultrasonic flaw detection of alloy steel rail, forged point rail, point, wing rail insert, and alloy steel section of welded wing rail		No defect (discontinuous) failing to meet the acceptance standards	One by one	A	√	√
10		Content of sulfur, phosphorus and hydrogen in alloy steel	S	≤0.015%	One by one	B	√	
			P	≤0.025%				
			H	≤0.000 15%				
11		Appearance of welds of welded wing rail	Straightness	Complying with Table 2 in TB/T 1632.2-2014	One by one	A	√	√
			Surface quality					
12		Welding of welded wing rail		Complying with TB/T 1632.2-2014	Once every five years	A	√	
13		Physical and mechanical properties of gluing materials		Complying with TB/T 2975	As per Article 6.2.1	A	√	
14		Appearance of forged point rail and wing rail insert (oxide scale)		Top: without oxide scale	One by one	C	√	√
				Bottom: ≤1 mm × 50 mm (Depth × length), interval ≥200 mm				
				Side: ≤0.5 mm × 50 mm (Depth × length), interval ≥150 mm				
15		Tolerance for the length from the actual point of crossing to the starting point of binding surface of the heel rail		±2 mm	One by one	B	√	√
16		Tolerance for the cross section height of point rail head at 20 mm-50 mm		±0.5 mm	One by one	B	√	√
17		Tolerance foreach cross section width of point rail		±0.5 mm	One by one	B	√	√
18		Inclination of rail end face (horizontal and vertical)		<1.0 mm	One by one	B	√	√
19		Tolerance for the length of long point rail, short point rail and heel rail		±2 mm	One by one	C	√	√
20		Tolerance for the base width of long point rail, short point rail and heel rail		−2 mm, 0 mm	One by one	C	√	√

Table A Basic Items for Crossing Inspection(continued)

S/N	Inspection item		Requirements	Frequency	Classification of inspection item	Type test	Delivery inspection
21	Rail components and parts	Tolerance for the length of wing rail	±6 mm	One by one	C	√	√
22		Tolerance for the height of wing rail	±0.5 mm	One by one	C	√	√
23		Tolerance for the width of wing rail base	−2 mm, 0 mm	One by one	C	√	√
24		Tolerance for the head width of long point rail, short point rail, heel rail and wing rail (except for the bending point)	±0.5 mm	One by one	B	√	√
25		Tolerance for the profile of the section rail head of long and short point rails with a rail head width of 35 mm, 50 mm and 71 mm (or 70 mm)	⩽0.3 mm	One by one	B	√	√
26		Tolerance for the length of wing rail insert	−2 mm, 0 mm	One by one	C	√	√
27		Tolerance for the width of each control section of wing rail insert	±0.5 mm	One by one	B	√	√
28		Tolerance for the height of each control section of wing rail insert	0 mm, +1.0 mm	One by one	B	√	√
29		Tolerance for the section thickness of wing rail insert	±0.5 mm	One by one	B	√	√
30		Asymmetry for the machined surface of point rail of straight crossing	⩽0.5 mm	One by one	C	√	√
31		Indentation depth formed in bending	⩽0.5 mm	One by one	B	√	√
32		Surface roughness	*Ra*25 μm	One by one	B	√	√
33		Tolerance of bolt hole diameter	0 mm, +1 mm	One by one	C	√	√
34		Tolerance for the upper and lower positions of bolt hole center	±1.0 mm	One by one	C	√	√
35		Tolerance for the distance from the center of the joint bolt hole to the rail end	±1.0 mm	One by one	B	√	√

Table A Basic Items for Crossing Inspection(continued)

S/N	Inspection item			Requirements	Frequency	Classification of inspection item	Type test	Delivery inspection
36	Rail components and parts	Tolerance of the center to center distance of two adjacent bolt holes	In To be assembled into one piece	±1.0 mm	One by one	C	√	√
37			Not to be assembled into one piece	±2.0 mm				
38		Tolerance of the center to center distance between two bolt holes with the furthest distance		±3.0 mm	One by one	B	√	√
39		Tolerance for the length of base plate		±3 mm	One by one	C	√	√
40		Tolerance for the width of base plate		±2 mm	One by one	B	√	√
41		Tolerance for the thickness of base plate		±0.5 mm	One by one	B	√	√
42		Tolerance of the distance from the gauge blocks mounted on both sides of the base plate to the running surface		±1.0 mm	One by one	B	√	√
43	Base Plate	Tolerance of the distance between two iron bases on the pad		±1.0 mm	One by one	B	√	√
44		Tolerance for the parallelism of two parallel iron bases on the base plate		1.0 mm	One by one	B	√	√
45		Tolerance for the rail cant of base plate		±1∶320	One by one	B	√	√
46		Chamfering of base plate bolt holes		1.0 mm×45°-1.5 mm×45°	One by one	B	√	√
47		Appearance of base plate welds		The welds shall be uniform and compact, and no residual overlap, welding slag, flash and burr are all owed on each surface of the base plate	One by one	B	√	√
48	Complete set of crossings	Binding gap between long and short point rail heads		<0.5 mm	One by one	B	√	√
49		Binding surface gap between the heel rail head and the point rail		<0.5 mm	One by one	B	√	√

Table A　Basic Items for Crossing Inspection(continued)

S/N	Inspection item		Requirements	Frequency	Classification of inspection item	Type test	Delivery inspection
50	Complete set of crossings	Gap between long and short point rail bases	≥1.0 mm	One by one	B	√	√
51		Gap between wing rail base and point rail base	≥2.0 mm	One by one	B	√	√
52		Gap between wing rail head and insert as well as between point rail and heel rail	<0.5 mm	One by one	B	√	√
53		Height difference (h) between the wing rail and each control section at the front end of the insert	$^{0}_{-1.0}$ mm	One by one	B	√	√
54		Gap between insert and wing rail base	<0.5 mm	One by one	B	√	√
55		Tolerance for the total length of crossings	±4 mm	One by one	C	√	√
56		Tolerance of the length from the crossing toe end and heel end to the theoretical point of crossing	±2 mm	One by one	C	√	√
57		Tolerance of toe opening	±2 mm	One by one	B	√	√
58		Tolerance of heel opening	±2 mm	One by one	B	√	√
59		Tolerance of throat width	$^{+2.0}_{0}$ mm	One by one	B	√	√
60		Tolerance of flangeway width in the buffer section at the back end of wing rail	±2.0 mm	One by one	B	√	√
61		Tolerance of flangeway width in the cross section of point rail with a rail head width of 20 mm and 50 mm	$^{+1.0}_{-0.5}$ mm	One by one	B	√	√
62		Tolerance for the height between the top of point rail to that of wing rail (or insert) in the cross section of the point rail with a rail head width of 20 mm and 50 mm (or 45 mm)	±0.5 mm	One by one	B	√	√
63		Tolerance for the offset of long and short point rails	±1.0 mm	One by one	B	√	√

Table A　Basic Items for Crossing Inspection(continued)

S/N	Inspection item		Requirements	Frequency	Classification of inspection item	Type test	Delivery inspection
64	Complete set of crossings	Tolerance for offset of each control section of crossing (curved crossing)	±1.0 mm	One by one	B	√	√
65		Narrow gap between spacing block and point rail, wing rail and heel rail, all of which are not glued	<0.5 mm	One by one	B	√	√
66		Gap between the wing rail and point rail base and the elastic pad or iron base plate	<0.5 mm	One by one	C	√	√
67		Tolerance of spacing between adjacent iron base plates	±5 mm	One by one	C	√	√
68		Tolerance for the torsional angle of rail cant	±1∶320	One by one	B	√	√
69		Straightness of running surface of crossing	<2.0 mm. There shall be no crossing within the section area of 50 mm from the throat to the point rail, or the curve shall be round and smooth	One by one	B	√	√
70		Chamfer of the binding surfaces formed by rail head and point rail and insert, etc.	1.5 mm×45°-2.5 mm×45°	One by one	C	√	√
71		Chamfer of rail end face, machined rail head and rail base	1.0 mm×45°-2.0 mm×45°	One by one	C	√	√
72		Chamfer of bolt holes for point rail, rail and insert	0.8 mm×45°-1.5 mm×45° or grounded	One by one	B	√	√
73		Torque of high-strength bolts	100%-110% of the design value	One by one	B	√	√
74		Flangeway depth	≥47 mm	One by one	A	√	√
75		Marks	Correct and complete	One by one	B	√	√